# TROPICAL BIRDS

# TROPICAL BIRDS

**Tom Jackson**

amber
BOOKS

Published by Amber Books Ltd
United House
North Road
London
N7 9DP
United Kingdom
www.amberbooks.co.uk
Instagram: amberbooksltd
Facebook: amberbooks
Twitter: @amberbooks
Pinterest: amberbooksltd



ISBN: 978-1-83886-196-4

Project Editor: Michael Spilling
Designer: Keren Harragan
Picture Research: Justin Willsdon

Printed in Malaysia

# Contents

# Introduction

If you want to enjoy what the world of birds has to offer, there is no better place to start than in the tropics. There are a little more than 10,000 species of birds in the world, and almost 90 per cent of them will spend time in the tropical regions of Earth in a given year. The tropics refer to a belt of the Earth's surface that straddles the Equator, where the four seasons of the temperate world do not apply. Instead, these areas may experience wet and dry spells but it is always hot, hot, hot!

Many birds, especially songbirds, visit the tropics in winter to escape harsher weather north and south. Many other bird families are thoroughly entrenched as tropical residents. In deserts and arid savannahs, these birds opt for drab colours to blend in with their bleak, open habitats. However, in tropical forests it is easy to hide, perhaps too easy! As a result many tropical birds have evolved a technicolour plumage as a way of getting noticed among the dense foliage. Let's take a tour of the most vibrant, eye-catching and simply beautiful tropical birds the world has to offer.

ABOVE:
A breeding pair of sun conures, a small South American parrot, share quality time together.

OPPOSITE:
One of the largest parrots out there, this red-and-green macaw needs to spend time preening and maintaining its wonderfully colourful feathers.

# Parrots and Cuckoos

Few birds have the appeal of parrots, with their bright plumage and cheeky chatter. These engaging birds are a mainstay of tropical habitats the world over, especially so in the Southern Hemisphere. There are about 330 species of parrot, forming the bird order Psittaciformes. Within that, the birds are further divided into the macaws and amazons of South and Central America, the cockatoos of Australia and the Indonesian Archipelago, and the lorikeets (or lories), which also live mainly in Australasia, where they specialise in a liquid diet. True parrots populate the Americas, Africa and Asia. Some readers will be familiar with the term parakeet, and this is applied to a wide range of smaller parrot species, often unrelated.

Traditionally, the parrots are seen as a close cousin to the cuckoos – both share some morphological features and are renowned for their distinctive, although entirely different, vocalizations. The cuckoos belong to the 150 or so species in the Cuculiformes order, along with turacos and the bizarre hoatzin of the Amazon jungles. Genetic analysis has recently rewritten much of the bird family tree and quashes any close link between the two groups, placing parrots more closely related to falcons!

OPPOSITE:
**Double-eyed fig-parrot**
The smallest parrot in Australia – although more common in New Guinea – this little bird is named after the blue spot between the eye and nostril, roughly the same size as the eye. This is the male of the species; the female lacks the red facial markings.

ABOVE TOP:
**Red-winged parrot**
A common bird in northern and eastern Australia, this species is seen in flocks close to bodies of water, but is otherwise seen in smaller groups feeding in grassland, shrubs, woodlands and along coastal forests.

ABOVE BOTTOM:
**Rose-ringed parakeet**
Native to Central Africa and South Asia this parakeet has established feral populations across northern Europe, confounding ecologists as to how these tropical birds are able to thrive, finding food and breeding sites, in these colder and harsher climates.

RIGHT:
**Splendid parrot**
A female and male splendid parrot sit side by side. This Australian species is often called the scarlet-breasted parrot due to the brighter plumage of the male on the right.

LEFT:

**Eclectus parrot**

The red plumage shows that this is a female of this species, from Australia and the Melanesian islands. Unusually, the male is less eye-catching and is largely green.

ABOVE:

**Pesquet's parrot**

The unusually long and viscously hooked bill of this Brazilian species earns it the alternative names of vulturine parrot or Dracula parrot. Despite the blood-thirsty look, this bird eats fruit, especially figs.

**Australian king parrot**
The red breast shows that this is a male of this Australian species, which is seen living along the Queensland and Victoria coast.

OPPOSITE:

**Kea**

The kea is a highly unusual parrot from New Zealand. While most parrots are fruit eaters, this alpine species is a scavenger, feeding on carrion as well as insects, seeds and leaves.

ABOVE:

**Burrowing parrots**

As the name suggests, this species, from Argentina and Chile, digs burrows into the dry loose soil of cliffs. The burrows reach 3m (10ft) in depth and connect with others to make a parrot warren where the chicks are raised.

LEFT:

**Red fan parrot**

The long, red-tipped feathers on the neck can be raised to make an colourful halo around the head, which is meant to startle and impress. The red fan parrot lives in the wider Amazon region.

OPPOSITE:
**Vernal hanging parrot**
This Asian bird is named after the way it often perches head down to reach dangling fruits.

ABOVE:
**Cuban amazon**
Found in the Bahamas and Cayman Islands as well as Cuba, this parrot has a red throat patch.

LEFT:
**Black-capped lory**
Found in New Guinea, this parrot has a wider diet than most, feasting on nectar, pollen and insects, as well as soft fruit.

ABOVE:
**Black lory**
Living in West Papua, the western side of New Guinea, this parrot has an unusually dark and inconspicuous plumage.

LEFT:
**Goldie's lorikeet**
Lorikeets (and lories), such as this New Guinean species, have feathery tongues inside delicate bills. While other parrots gouge out fruits from hard skins, lorikeets lick up food like pollen and nectar, and the juices of softer fruits.

ABOVE:
**Chattering lory**
Found only in North Maluku, a remote island of Indonesia, this parrot is considered vulnerable to extinction due to birds being captured from the wild to be sold as pets.

OPPOSITE:
**Scaly-breasted lorikeets**
The pattern of yellow tips on green feathers gives this Australian woodland bird's plumage a tile pattern that resembles the scales of a reptile.

ABOVE:
**Red lory**
Unusually there is no clear distinction between the male and female of this Indonesian species.

**Budgerigar**
One of the smallest members of the parrot family, this little bird is native to Australia but is now widespread as a popular pet. The source of the name is not clear but it is possibly a corruption of an Indigenous Australian phrase meaning 'good bird'. It has been proposed that flocks of wild budgies gathered in places with a good supply of water, and were thus a welcome sight for local people seeking food and drink.

LEFT:

**Crimson rosella**

This species was first described in the 1780s from specimens returned to Europe from its native Australia by the famous English naturalist, Joseph Banks.

ABOVE TOP:

**Eastern rosella**

This is one of six rosella species all living in Australia. The name is ultimately derived from Rose Hill, New South Wales, where the eastern rosella was first encountered by Europeans.

ABOVE BOTTOM:

**Western rosella**

Found in the woodlands of Western Australia, this is the smallest rosella species at 26cm (10in) long.

ABOVE AND OPPOSITE BOTTOM:
**Yellow-collared lovebirds**
The dark face of this little African parrot has earned it the alternative name of masked lovebird.

OPPOSITE TOP LEFT:
**Black-cheeked lovebird**
Living in dry woodlands in southern Africa, this little bird is becoming increasingly endangered due to habitat loss.

OPPOSITE TOP RIGHT:
**Fischer's lovebird**
The lovebirds are small African parrots that are so-named because they are often seen in pairs.

LEFT:
**Plum-headed parakeet**
Only the male of this Indian species has the eponymous plum-coloured head. The female has a grey head.

ABOVE TOP:
**Plain parakeet**
Only the pink beak and feet break up the plain green colouring of this aptly named parrot, which is a common sight in the urban areas of southern Brazil.

ABOVE BOTTOM:
**Grey-headed lovebird**
As a resident of Madagascar, this is the only lovebird that lives outside the African mainland.

**Sun conure**
Also known as the sun parakeet, this gregarious bird lives in large flocks in the forests of Venezuela and Guinea Highlands.

ABOVE:
**Harlequin macaw**
This colourful bird is a hybrid between a blue and gold macaw and a green-winged macaw.

RIGHT:
**Red and green macaws**
The long tails typical of the macaws of the Americas are clearly in view as this flock of birds soars over the treetops.

**Military macaw**
The red, green and blue colouring of this big parrot, native to Central and South America, reminded naturalists of the parade uniforms of Latin American soldiers.

OPPOSITE:

**Chestnut-fronted macaw**

Largely green with a flash of red on the sides of the wings, this species is also called the severe macaw. That name comes from the way these gregarious birds become belligerent in early adulthood.

LEFT TOP:

**Blue and yellow macaw**

Growing to 80cm (31in) long from beak to the tip of its long tail feathers, this vibrant macaw can live for 70 years. It is a widespread species that lives beyond the jungle in savannah and woodland.

LEFT MIDDLE:

**Military macaw**

Living in the forests of the Andean foothills and rugged hillsides elsewhere in northern South America, and further north in Mexico, this species nests high in the trees and lives in large flocks. At dawn it sets off to feed on fruits and leaves.

LEFT BOTTOM:

**Scarlet macaw**

Widespread across the Amazon and Central America, this big macaw lives in rainforests. Its tail makes up more than half of the total body length, a greater proportion than any other parrot.

LEFT:
**Scarlet macaw**
Despite the name, the scarlet macaw's plumage is filled with yellows, greens and blues.

ABOVE TOP:
**Red and green macaw**
This species can be easily mistaken for other macaws, but the dashed red rings around the eye are a distinguishing feature.

ABOVE BOTTOM:
**Blue and yellow macaw**
Stripes of black surround the eye on the otherwise white face of this large macaw.

ABOVE TOP AND BOTTOM:
**Cockatiel**
Also known as the quarrion or weiro, this small parrot is found across Australia, aside from the driest part of the interior. The wild type is grey but there are many more colourful domestic breeds kept as pets.

RIGHT:
**Palm cockatoo**
The sleek crest and huge beak have earned this Australian parrot the alternative name of goliath cockatoo. The cockatoos, all found in and around Australia, are distinguished from other parrots by their feathery crests.

LEFT:
**Pink cockatoo**
A pink cockatoo shows off its crest as it takes a dip in a desert pool.

ABOVE TOP:
**Galah**
This is the most common of all cockatoos. Large flocks are seen as far afield as the Philippines and Tasmania.

ABOVE BOTTOM:
**Gang gang cockatoo**
A male gang gang eats some berries in the alpine woodlands of Australia. This species is found in cooler regions of that continent.

ABOVE:

**Sulphur-crested cockatoo**

Ranging from New Guinea, and along the northern and eastern regions of Australia to Tasmania, this parrot is famed for making loud calls in its forest habitat.

RIGHT:

**Glossy black cockatoo**

A female glossy black cockatoo on a perch. This bird is a specialist consumer of pine cones

RIGHT MIDDLE:

**Yellow-tailed black cockatoo**

The yellow bars on the underside of the tail feathers show where this bird got its name.

FAR RIGHT:

**White cockatoo**

This species from Indonesia is also called the umbrella cockatoo because of the way the crest unfurls.

OPPOSITE LEFT:
**Bare-faced go-away bird**
Named after the sound of its call (and apparently bare face) this East African species is a kind of turaco, or banana-eater. These birds are traditionally linked to cuckoos and parrots because they all have feet with two forward-facing toes and two that face backwards, creating a strongly grasping foot.

OPPOSITE RIGHT TOP:
**Purple-crested turaco**
This southern African species eats fruits and lives in moist rainforests.

OPPOSITE RIGHT BOTTOM:
**Western plantain-eater**
This is the West African species of the plantain eaters. Despite its name it is more likely to eat figs, seeds and leaves.

LEFT TOP:
**Hoatzin**
Found only in the Amazon, this bizarre bird has claws on its wings, which the younger fledglings use to cling to branches as they clamber up trees and learn to fly.

LEFT BOTTOM:
**Indian cuckoo**
In common with many cuckoos, this Asian species does not build its own nests. Instead, it is a brood parasite, meaning it leaves its eggs in the nest of other species. The cuckoo chick muscles out any of the resident chicks and is raised by the duped foster parents – not by its actual parents.

OPPOSITE:
**Asian koel**
This cuckoo species is a brood parasite of the crow, shrike and myna bird.

LEFT TOP:
**Asian emerald cuckoo**
Widely spread across South and Southeast Asia, this vibrant species parasitises the crimson sunbird and little spiderhunter.

LEFT MIDDLE:
**Guira**
A South American cuckoo species, the guira is not a brood parasite. Instead it lives in large flocks and often lays eggs in communal nests. After hatching, chicks may push out unhatched eggs to ensure they have space in the nest. This is also something parasitic cuckoos do.

LEFT BOTTOM:
**Greater coucal**
Also called the crow pheasant this cuckoo species lives in South and Southeast Asia. It is not a parasite. The large bird hunts on the ground for insects and reptiles.

ABOVE TOP:
**Roadrunner**
Famed for its appearance in cartoons, this relative of the cuckoo is so-named because it can run fast and tends to move in straight lines across flat, arid habitats (roads are ideal). This smaller species has a distinctive blue patch around the eye.

ABOVE BOTTOM AND RIGHT:
**Greater roadrunner**
The larger of the two roadrunner species, this bird uses its speed to chase prey such as snakes. While running the bird sticks its tail out flat and lowers its head. Both roadrunner species live in Mexico and the south-western United States.

OPPOSITE:
**Crested coua**
This impressive cuckoo with a punk hair style and garish eye colouring is a native of Madagascar.

LEFT:
**Chestnut-breasted malkoha**
This large non-parasitic cuckoo lives in Southeast Asia, as far south as Java.

# Toucans and Woodpeckers

Famed for the chunky-looking colourful beak built to attract attention in the dense jungle habitat, the toucans are surprisingly agile in the air. These American birds, which are also named toucanets and aracaris, are the most flamboyant members of the Piciformes order, but nevertheless get a lot of competition from the 450 species in that group.

Along with the fruit-and-nut eating toucans, the order contains their close relatives, the barbets. These birds are smaller and spread more widely across the globe than the toucans, but share a hint of the same chunky beak. The barbets are fruit-eaters, too – although some glean insects as well – yet the sturdy beak is also used for digging a nest hole in dead wood, a feature that is shared with more famous Piciformes members, the woodpeckers. The woodpeckers are a very widespread group, often associated with the northern forests and woodlands of colder, northern regions. However, many species also make their homes in tropical forests. The final Piciformes birds that require a mention are the honeyguides, mostly of Africa, which feast on beeswax, and have a cunning tactic for accessing this food source.

OPPOSITE:
**Channel-billed toucan**
Despite its large size – making up a third of its total body length – the immense bill, or beak, is very lightweight thanks to a hollow, honeycomb structure inside.

ABOVE TOP:
**Curl-crested aracari**
The bill has a serrated edge for gripping food items in flight and slicing into fruits.

ABOVE BOTTOM:
**Pale-mandibled aracari**
Generally agreed to be a subspecies of the collared aracari, which lives long the coastal forests of northern South America, this particular bird lives in the western parts of Ecuador and Peru.

RIGHT:
**Keel-billed toucan**
This vibrant species is typical of toucans in that it will mainly eat fruits plucked from forest trees, but also insects and small lizards if the opportunity arises.

OPPOSITE TOP:

**Fiery-billed aracari**

Hailing from the Pacific coastal forests of Costa Rica and Panama, this bird is typical of the aracaris in that it roosts in a small flock. Other toucans are largely solitary.

OPPOSITE BOTTOM:

**Golden-collared toucanet**

As the name suggests, these birds are smaller than other toucans, being about two-thirds of the size of other adult members of the family.

BELOW:

**Toco toucan**

At 65cm (26in) long, this widespread South American species is the largest of all toucans. Despite the stout-looking body, all toucans are agile fliers with wide wings ideal for short, slow-speed flights.

**Plate-billed mountain toucan**
The mountains referred to here are the western foothills of the Ecuadorian Andes. The red-brown eye shows that is a colour morph from the northern part of the range. To the south, this species has greener eyes.

OPPOSITE:

**Greater honeyguide**
The only species known to use it calls to lead humans to wild beehives, this southern African spices has the scientific name *Indicator indicator*. The human then breaks up the beehive to get at the honey, while the bird eats the wax and larvae.

LEFT TOP:

**Scaly-throated honeyguide**
Widespread throughout the tropics of Africa, it is uncertain if this species lives up to its name. The birds belong to the family Indicatoridae, and most honeyguides are in the genus *Indicator*. However, only the greater honeyguide is proven to actually guide people to bee hives.

LEFT MIDDLE:

**Malaysian honeyguide**
This Southeast Asian species feasts on bees and wasps, and has a distinctive call that sounds like a cat's meow.

LEFT BOTTOM:

**Spotted honeyguide**
Not a lot is known about this African species, but all honeyguides studied to date are brood parasites like cuckoos. As a member of the *Indicator* genus, this species probably lays its egg in the nest of its close relatives, the woodpeckers and barbets. A honeyguide hatchling cracks the eggs of its nest-mates to ensure they are the only chick in the nest.

LEFT:
**Brown-headed barbet**
This fruit-eating Asian species is happy to share its home with humans so it is a frequent visitor to city parks, orchards and other agricultural areas.

BELOW TOP:
**Lineated barbet**
This species lives in the forests of Bangladesh and the surrounding areas drained by the mighty Brahmaputra River.

BELOW BOTTOM:
**Blue-eared barbet**
Found across Southeast Asia, this species is named after the blue patch behind the eye. This one is a male with a dark band on the forehead.

OPPOSITE:

**Blue-throated barbet**

This colourful bird is a frequent visitor to gardens and parks in the Malay peninsula and Sumatra.

LEFT TOP:

**Fire-tufted barbet**

Both sexes have a beard-like tufts of feathers under the beak – which earns the whole group the name 'barbet' – but only the males of this species have the orange-tipped tuft on the upper beak.

LEFT MIDDLE:

**Great barbet**

At 35cm (14in) long, this is the largest barbet species. It lives across India and Southeast Asia, and is especially common in the forested foothills of the Himalayas. In common with other barbets, it eats fruits and insects.

LEFT BOTTOM:

**Taiwan barbet**

The Chinese name for this species is 'five-coloured bird'. Like their woodpecker cousins, the barbets make their nests in tree holes, using their sturdy beaks to bore out living space in the wood.

LEFT:
**Gilded flicker**
With is distinctive golden or gilded wings, this woodpecker species is limited to the deserts of the southwest United States, and the region around Mexico's Gulf of California. It usually builds its nests in saguaro cacti.

ABOVE TOP:
**Red-bellied woodpecker**
This American woodpecker qualifies as tropical because it lives in the forested swamps of Florida and the Deep South (as well as the pine forests of Canada). The red cap extends further forward in this male specimen.

ABOVE BOTTOM:
**Greater yellownape**
Named for the yellow colouring on the back of the head, this species of woodpecker lives in southern China, Indochina and Sumatra.

OPPOSITE:

**Red-headed woodpecker**

A medium-sized woodpecker that is widely spread over eastern North America, but extends into sultry forests of Florida and the Gulf of Mexico.

LEFT:

**Pileated woodpecker**

A male pileated woodpecker turns up with lunch for a pair of chicks. The female of the species lacks the red check patch. This is the second-largest woodpecker in the United States, found as far south as Florida. That state is the last refuge of the largest species, the ivory-billed woodpecker, which has not been seen in the wild since 1944.

ALL PHOTOGRAPHS:
**Acorn woodpecker**
These North American woodpeckers create a 'granary' in tree trunks to store acorns. Pictured here are males (with a bigger red cap than the females) stocking acorns inside small holes chiselled into dead wood. These unusual birds live in the foothills of the Rockies, and further ranges south through Central America.

OPPOSITE:

**Magellanic woodpecker**

Although not strictly tropical, this enormous Patagonian woodpecker, 40cm (16in) tall, has an angry red head plumage.

LEFT:

**Yellow-tufted woodpecker**

This woodpecker lives in the Amazon rainforest and similar habitats across the tropical parts of South America.

BELOW:

**Greater flameback**

A pair of these woodpeckers dig for termites and grubs inside the a tree trunk. This species lives in South and East Asia.

ABOVE AND RIGHT:
**Black-cheeked woodpecker**
This species lives in Mexico and much of Central America. It supplements its diet of insects with fruits and nectar. It digs a nest high in the canopy inside a dead branch.

# Hornbills and Relatives

At first glace the hornbills seem to be a close relative of the toucans, and indeed both groups of bird use large beaks to handle a range of tough fruit foods – as well as an eclectic animal diet. However, while the toucans are a group found in tropical America, the hornbills are Old World birds living in Africa, mainland Asia and the Indonesian Archipelago. Many of the 45 hornbill species are easily distinguished by the hollow ornamental casque on the top of the beak. The hornbills occupy the Bucerotiformes order, which they share with the hoopoes, a group with more delicate beaks to go with their ornamental headgear, which takes the form of a crown of feathers.

A linked order is the Coraciiformes with members that are also similarly characterized by a long, capable beak. With occupants more suited to hunting rather a fruit diet some authorities place this order closer to the woodpeckers and their allies than to the hornbills. The Coraciiformes includes the 86 species of kingfishers, which are a cosmopolitan bunch living across all continents, including Australia, where they are better known as kookaburras. And most of the kingfishers live far from rivers and do not fish at all! Also in the Coraciiformes are the 24 species of bee-eaters, colourful and agile birds that are found in all tropical areas outside of the Americas. They have an Old World distribution in common with the 17 species of rollers, which were once grouped with the hoopoes by older systems. Finally, the Coraciiformes includes eight species of motmots, small insect-eaters living in Central and South America, and five species of todies, which have a similar lifestyle and are confined to the Carribean.

OPPOSITE:
**Oriental pied hornbill**
This large hornbill from the forests of Southeast Asia and Indonesia mostly eats fruits such as figs, but also insects, lizards and even the occasional bat!

OPPOSITE TOP LEFT AND RIGHT:
**Rhinoceros hornbill**
This big bird is named for the impressive horn on its head. The animal grows to 90cm (35in) long.

RIGHT AND BELOW:
**Knobbed hornbill**
The female below has a muted yellow casque, while the male, right, has a dark red one.

**Oriental pied hornbill**
This close-up shows the horny nature of the bill or beak. The sturdy bill is used for fighting as well as handling foods.

RIGHT TOP:
**Rufous hornbill**
Local people in the Philippines call this bird the 'clock of the mountains', thanks to its booming call.

RIGHT MIDDLE:
**Southern yellow-billed hornbill**
Found in southern Africa, this hornbill is found mostly on the ground. It forages for insects and other small animals among the shrubs, along with seeds and fruits.

RIGHT BOTTOM:
**Wreathed hornbill**
Named for the distinctive bars on the upper beak, this one is the male of the Asian species, distinguised by its yellow throat pouch. Females have a blue one.

OPPOSITE TOP:
**Sumba hornbill**
Despite their large size and heavy beaks, hornbills, such as this one from the Indonesian island of Sumba, are powerful fliers.

OPPOSITE BOTTOM:
**Great hornbill**
Hailing from India and Southeast Asia, this is the heaviest hornbill species, weighing 7kg (15lb).

LEFT:
**Belted kingfisher**
A water kingfisher from North America, this species moves north to Canada to breed but spends the winter in Mexico and much of Central America.

ABOVE:
**Sacred kingfisher**
Found across much of Australia, New Zealand and the Pacific islands, this kingfisher lives in mangroves and woodlands where it feeds on insects, frogs and other small animals. Polynesian people believe the bird controls the waves, hence the sacred moniker.

ABOVE TOP AND BOTTOM:

**Woodland kingfisher**

As the name suggests, this African species makes its life in the trees, hunting for insects from a branch.

RIGHT:

**Oriental dwarf kingfisher**

Aptly named, this Asian species is typically about half the size of its cousins at about 14cm (6in) long.

LEFT:
**Giant kingfisher**
The largest kingfisher in Africa, this bird is 46cm (18in) long and is found all over the continent south of the Sahara. It hunts for fish, frogs and crabs in rivers.

ABOVE TOP AND BOTTOM:
**Pied kingfisher**
Named for its black and white colouring, this species is one of the most widespread tropical kingfishers. It hunts in rivers across Africa and Asia, from the Cape of Good Hope to China.

LEFT:
**Blue-winged kookaburra**
This species from New Guinea and northern Australia is a ground hunter, foraging for a wide range of prey. In moist habitats it relies on frogs and lizards but in drier conditions it eats insects, worms and scorpions.

BELOW TOP:
**Rufous-bellied kookaburra**
Smaller than most kookaburras, this brightly coloured kingfisher lives in the lowland forests of New Guinea.

BELOW BOTTOM:
**Spangled kookaburra**
This little species lives on the Aru Islands of eastern Indonesia and in southern New Guinea. The females are larger than the males but both sexes have the same plumage.

ALL PHOTOGRAPHS:

**Laughing kookaburra**

This famous Australian bird is the largest of all the kingfishers, at 47cm (19in) long and weighing 400g (14oz). It lives in the eastern region of the continent, with a small population in the Perth area to the west. The name kookaburra, which is applied to other Australian kingfishers, comes from this ground hunter's laughing call.

OPPOSITE TOP LEFT:
**Andean motmot**
Also known as the highland motmot, this species lives in the mountain forests of Colombia and Bolivia, often near streams. It is suprisingly large, growing to 46cm (18in) including the long tail.

OPPOSITE TOP RIGHT:
**Blue-throated motmot**
This species lives in the humid montain forests of Central America. This picture shows off the racquet-tipped tail feathers that are typical of all motmots.

OPPOSITE BOTTOM LEFT:
**Keel-billed motmot**
This endangered species from the mountains of Belize, Costa Rica, Guatemala, Honduras, Nicaragua, and southern Mexico snatches insects on the wing.

OPPOSITE BOTTOM RIGHT:
**Rufous motmot**
The second-largest motmot at 47cm (19in) long, this species is widepread across the southern part of Central America and much of the western Amazon Basin, east of the Andes.

LEFT:
**Blue-crowned motmot**
Found in the dense forests along the coast of eastern Mexico, this bird has an owl-like call. They are primarily insect-eaters but also eat a few fruits.

OVERLEAF:
**Puerto Rican tody**
Thought to be descended from Jamaican todies blown to the island during hurricanes, this little bird survives almost entirely on insects, which it gleans from the leaves of shrubs and trees.

ABOVE TOP:
**European bee-eater**
Misnamed perhaps because this bird is far more common beyond its populations in southern Europe. It also lives in the Magreb region of North Africa and the Middle East.

ABOVE BOTTOM:
**Böhms bee-eater**
Named for a German naturalist, this species lives in Central and southern Africa.

RIGHT:
**Southern carmine bee-eater**
Occupying parts of southern Africa north of the Kalahari and Cape regions, this species hunts close to large animals that disturb insects which the birds then snatch mid-air.

**Blue-throated bee-eater**
After spending winter in the Malay Peninsula and Indochina, this bird migrates north in the summer to breed in southern China, often crossing large areas of ocean to get there.

**Red-throated bee-eater**
This widespread African species lives in large colonies. It prefers open habitats and is a common sight on farmland.

**Northern carmine bee-eater**
Seen across much of equatorial Africa from the Atlantic coast to the Horn of Africa, the carmine bee-eater is a colonial species. It lives in communal nests dug into dry, sandy cliffs.

RIGHT:

**Common hoopoe**

Living across the warmer parts of Europe and Central Asia, all the way to eastern China, this species is typical of the hoopoes in that it has a distinctive spiked crown on its head. The long bill is for probing in soil for food.

OPPOSITE RIGHT TOP:

**Madagascar hoopoe**

The name hoopoe comes from the oop-oop-oop calls it gives.

OPPOSITE RIGHT MIDDLE AND BOTTOM:

**African hoopoe**

The crown is used for display and to make the bird seem larger during periods of danger. In flight, the crown is folded flat.

RIGHT AND OPPOSITE:
**Lilac-breasted roller**
Like its cousins the bee-eaters, this African roller has adopted a strategy of feeding near to large grazing animals, which disturb insects as they walk along. On the right, the bird has grabbed an insect escaping a zebra's hooves and is now sitting on its companion's back to enjoy the meal.

BELOW:
**Indian roller**
As well as living in India and Pakistan this species inhabits the coasts of the Persian Gulf.

OPPOSITE:
**Blue-bellied roller**
Living in a band of equatorial Africa, this bird is seen in open areas in and among tropical forests and woodlands.

LEFT TOP AND MIDDLE:
**Broad-billed roller**
This African species lives in woodland, often near a stream or river. In the dry season, the bird migrates to lusher equatorial regions.

LEFT BOTTOM:
**European roller**
Despite the name, these birds also live across the Middle East and many will migrate south over the Sahara to spend the winter in Africa.

# Hummingbirds and Swifts

With a flash of iridescence and jaw-dropping aerial ability, the hummingbirds are an amazing sight as they flit between heavy blooms in lush tropical habitats. They are resident in North and South America, comprising 315 species. The birds are specialist feeders that target nectar (and sometimes small insects), which they acquire while hovering in midair. They beat their small triangular wings more than 50 times a second, so fast that the feathers form a humming blur of colour. The birds adjust their postion very precisely as they reach into the flower with their long beaks, and lick up food with an equally long tongue. The hummingbirds are the larger component of the Apodiformes order, which they share with the 74 species of swift. The swifts are also renowned for their flying skill, and famously some species spend almost their entire lives in the air once they have fledged. They even mate on the wing!

Also covered in this chapter are the trogons. This small group of American tropical birds have no particular close link to the hummingbirds but are included due to their similarly eye-catching plumage.

OPPOSITE:
**Sparkling violetear**
There is no need to explain from where this incredible Andean species got its name.

**Fiery-throated hummingbird**
This pair of minute birds, just 10cm (4in) from head to tail tip, can be found in the high mountain forests of Costa Rica and Panama.

ABOVE TOP AND BOTTOM:
**Green hermit**
This bird is from the mountain forests of Costa Rica and the northern Andes range. Hermits typically have a curved beak.

RIGHT:
**Bee hummingbird**
The bee hummingbird of Cuba is the smallest bird in the world at only 6cm (2.4in) long.

**Calliope hummingbird**
After wintering in Central America, this species migrates to the Pacifc forests of North America to breed. It builds a tiny cup-shaped nest from spiders' silk, strands of lichen, and moss. Hummingbird eggs are pea-sized and the smallest of any bird, yet they are still large in comparison to the mother's body size.

RIGHT:

**Saw-billed hermit**

Hailing from the Atlantic forests along the coast of south-eastern Brazil, this bird is classified as a hermit despite not showing the typical curved bill.

BOTTOM:

**Marvelous spatuletail**

Endemic to Peru, this bird is named after the shape of the impressive trailing tail feathers.

OPPOSITE:

**Copper-rumped hummingbird**

Mostly found in Venezeula, this bird is a common sight in open habitats, including gardens.

ABOVE:
**Rufous hummingbird**
Migrating between the Pacific Northwest and Mexico, this bird has a distinctive rusty red colouring, including the sparkling gorget (neck plumage). The iridescence of the feathers is due to nanoscopic structures on the surface that reflect different wavelengths of light in different ways.

RIGHT:
**Tufted coquette**
Found in the forests of the Guinea Highlands, this hummingbird is one of the comparative few to show obvious sexual dimorphisms. Only the male has the impressive crests and plumes. However, most hummingbirds do exhibit size dimorphism, where the females are bigger than the males and have longer beaks so they can gather extra food to produce eggs.

LEFT AND ABOVE TOP:

**Great dusky swifts**

A flock of swifts swoop past an immense cataract of the Iguazu Falls in Argentina. The birds are known to fly through the water and perch on the cliffs behind.

ABOVE BOTTOM:

**Moustached tree swift**

The tree swifts are close relatives of true swifts. However, they have long forked tails instead of the curved swallow tail typical of true swifts. (Swallows are songbirds and only distantly related.)

RIGHT:
**Black-throated trogon**
Sometimes named the yellow-bellied trogon for obvious reasons, this species lives mostly in the Amazon region but is also seen in Central America.

OPPOSITE LEFT:
**Cuban trogon**
In common with other trogons this impressive species is a national bird, in this case obviously Cuba, where it is endemic.

OPPOSITE RIGHT TOP:
**Orange-breasted trogon**
This trogon is found in Thailand and Vietnam. Its natural habitats are subtropical or tropical moist lowland forests. It has a typical trogon diet of insects.

OPPOSITE RIGHT BOTTOM:
**Masked trogon**
Named for its bandit-like face covering, this one is a female with a brown breast instead of the green of the male.

LEFT:
**Red-headed trogon**
Trogons hunt for food using a technique called 'sally-gleaning'. The bird spots prey on a branch, flies over and snatches it in its beak without landing and then returns to its perch to eat.

ABOVE:
**Surucua trogon**
This species is from southern Brazil and eastern Paraguay. It is sometimes called the orange-bellied trogon.

**Resplendent quetzal**
The quetzal is the national animal of Guatemala and appears on its flag. Even the local currency is called the quetzal. All that attention is warranted. Both sexes of this trogon are brightly coloured, with the male having a red belly and an immensely long tail, making up two-thirds of the total body length.

# Ground Birds

Larger birds tend to spend long periods of time on the ground, searching for food on foot. Some are simply too big to fly, others use their wings to escape threats with a short flight to safety, while others still are migratory, and travel long distances by air back and forth to breeding grounds each year before typically returning to spend the winter in tropical areas.

It goes without saying that this collection of birds is a rag-tag bunch of unrelated groups. The largest ground birds of all belong the ratite group. This name refers to the shape of the breast bone or sternum of these birds, which is flat like a raft instead of being keeled like a boat. Flight muscles are attached to the sternum, and the flat bone shows us that ratites have only very small muscles here, actuating equally small wings. The ratites do not fly at all and include ostriches, emus and rheas, all residents of tropical parts of the world.

Other ground birds include the game birds, such as pheasants and turkeys, which scratch out a living on the forest floor, and have some of the most resplendent plumages of any birds. Additionally, the tropics are home to many large semi-aquatic birds, such as herons, egrets and flamingos, as well as storks, spoonbills and ibises, which can also have a more terrestrial lifestyle.

OPPOSITE:
**Northern cassowary**
The purpose of the hollow casque on the head of this ratite is unclear. It may indicate dominance, health or help the head push through the forest.

RIGHT AND OPPOSITE BOTTOM LEFT:
**Southern cassowary**
Standing 180cm (71in) tall this jungle ratite runs through the lowland forest of southern New Guinea. American soldiers serving in the jungles in World War II perpeturated the entirely unsubstantiated rumour that the casque could be used as a weapon capable of piercing the chest!

OPPOSITE TOP:
**Northern cassowary**
Of similar size to its southern relative, this bird lives on the other side of New Guinea. It has sharp claws and can deliver a slashing kick by jumping into the air.

OPPOSITE BOTTOM RIGHT:
**Dwarf cassowary**
Occupying the highlands inland between the ranges of its larger cousins, this cassowary is 125cm (49in) tall. In common with the other species, young birds do not have casques.

ALL PHOTOGRAPHS THESE PAGES:
**Emus**
Standing at 1.9m (75in) tall, this is by far the largest bird in Australia. The males build a grassy nest area surrounded by sticks, and after laying a few eggs – each weighing 0.5kg (1lb) – the female departs, leaving the male to brood the clutch and raise the chicks.

ALL PHOTOGRAPHS OVERLEAF:
**Ostrich**
Measured at 2.8m (110in) tall, the ostrich of Africa's grasslands is the world's largest bird, and incapable of flight. It can, however, run at 70 km/h (44mph) making it one of the fastest creatures on land. It needs to be fast because its primary predator is the cheetah, which is the fastest land animal of all. The small wings are used for display and for steering when running at full pelt.

ABOVE:
**Ostrich hatchlings**
The male birds, distinguished by their darker feathers, look after the hatchlings.

RIGHT:
**Ostrich eggs**
The eggs, incubated by the female in a nest scraped into sand, are the largest of any animal on Earth at 15cm (6in) long and weighing around 1.3kg (2.9lb) each.

ABOVE TOP AND BOTTOM:
**Darwin's rhea**
Only distantly related to the ostrich and emus, the rheas are South American ratites found on the pampas grasslands. This is the smaller species, growing to 100cm (39in) tall.

RIGHT:
**Greater rhea**
Standing at 140cm (55in), this is the largest American bird. In common with other flightless ratites, the rhea's feathers are shaggy and unsuited for flight.

**Crested guineafowl**
This ground bird lives in the forests and woodland of Africa, south of the Sahara. Most guineafowl species are African, and several have been introduced to other continents as gamebird and livestock.

ALL PHOTOGRAPHS:
**Vulturine guineafowl**
This is the largest guineafowl species, growing to 70cm (28in) long. It is a gregarious bird that lives in large flocks. It can fly but tends to run from threats.

OPPOSITE LEFT AND RIGHT TOP:
**Nicobar pigeon**
The beautiful bird is found on only one remote island in the Indian Ocean. It is the closest living relative of the extinct dodo, another Indian Ocean bird.

OPPOSITE LEFT AND RIGHT BOTTOM:
**Temminck's tragopan**
This incredible-looking pheasant lives in Central China and the bordering countries. The female is a cryptic brown, ideal for blending into the forest floor habitat, but males like these are out to impress.

LEFT:
**Blood pheasant**
Found in the eastern Himalayas, both sexes of the blood pheasant have distinctive red eyes and legs, but only the male has bright blue green plumage.

ALL PHOTOGRAPHS:
**Silver pheasant**
Only the males of this Southeast Asian species display the metallic grey plumage. There are large number of subspecies with variations in colouring.

RIGHT:
**Golden pheasant**
Also known as the rainbow pheasant, this species lives in western China. The male grows to about 1m (39in) long, including the tail.

OPPOSITE TOP:
**Great argus**
With its tail feathers folded, the male of this Southeast Asian gamebird looks similar to the drab female, but can get some attention by opening up its fan of long feathers. Like its cousin the peafowl, females judge males by the symmetry of their fan.

OPPOSITE MIDDLE:
**Elliot's pheasant**
Named for the American ornithologist Daniel Elliot, this bird lives in southern China.

OPPOSITE BOTTOM:
**Palawan peacock pheasant**
Peacocks and pheasants are a closely related group, but this bird from the Philippines does not fan out its tail like its namesake.

LEFT:

**Thick-billed green pigeon**

A far cry from its urban relative, the rock dove that flocks in our city streets, this fruit-eating pigeon lives in South Asia. This specimen is the male.

ABOVE:

**Victoria crowned pigeon**

Standing to a height of 80cm (31in) thanks to its impressive head crest, this pigeon from New Guinea is the world's largest. It can fly but mostly stalks slowly around the forest floor.

LEFT AND ABOVE TOP:
**Australian brushturkey**
Despite the name, this bird is not a close relative of the domestic turkey, which hails from the Americas. Nevertheless, the eggs of this ground bird were an important food for Aboriginal Australians.

ABOVE BOTTOM:
**Scrubfowl**
This bird is a megapode – as are the brushturkeys – which means it does not incubate eggs using its body heat, as is the norm for other birds. Instead, it buries eggs inside a nest mound made from decaying vegetation, just as crocodiles do.

LEFT:

**Black heron**

This species, seen in wetlands across Africa south of the Sahara, is often called the black egret. It spreads its wings to cast a shadow over shallow waters. Fish are attracted to the shaded area for the bird to snap up.

ABOVE TOP:

**Great blue heron**

This impressive North American fish-eater spends the winter in Mexico and Central America, before migrating all the way up to Hudson Bay in summer. An all-white population of this species is resident in the Florida Keys.

ABOVE BOTTOM:

**Green heron**

At about 40cm (16in) tall, this species is considerably smaller than is typical of herons. In common with other American species it overwinters in Central America. It is said to use bait, such as insects dropped on the water, to lure fish prey to the surface.

**Great egret**

This large white heron is one of the most widespread bird species on Earth. This specimen is from the eastern subspecies that lives across South and Southeast Asia, and islands of the Indo-Pacific, including Australia and New Zealand. The western range covers Africa and South America, with some birds migrating into North America and southern Europe to breed in summer.

OPPOSITE:

**Least bittern**

A year-round resident of eastern South America – and a migrant into eastern North America – this small heron lurks in the tall vegetation of swamps and marshes. It feeds on small animals living in shallow water.

LEFT:

**Cattle egret**

A widely ranging species found in all tropical areas. It is named for its habit of accompanying herds of cattle or other large herbivores, perhaps hitching a ride on their backs. As the big animals move through the grass, they disturb insects and other small animals that are targeted by the birds.

RIGHT TOP:
**Saddle-billed stork**
Named for the black band that separates the two red sections seen in both sexes, this stork lives in swamps in sub-Saharan Africa. It is more common in West Africa than further to the south.

RIGHT MIDDLE:
**White stork**
The proverbial stork, this species migrates each year from Africa to nest in Europe, and a smaller population moves between Kazakhstan and western India.

RIGHT BOTTOM:
**Asian openbill**
Living across the Indian subcontinent and the mainland of Southeast Asia, this stork is named after the way that the beak always has a gap between the upper and lower beak, which is curved and slightly twisted. This unusual setup is used to remove snails from their shells.

OPPOSITE:
**Painted stork**
Mostly found in India, this stork feeds by swishing its beak through the water feeling for small fish, which are duly snapped up.

**Black stork**
Wintering in southern Africa, this large stork makes one of the world's most epic migrations as it tracks two main routes north in spring. After reaching North Africa, these big birds must cross the ocean. However, they need to ride warm thermals to stay aloft and so cannot fly for long distances over cold seawater. As a result, all the birds enter western Europe via the Straits of Gibralter and reach their eastern breeding grounds by crossing the Dardenelles.

**Roseate spoonbill**

With a small population in Florida and the Caribbean, most of these wading birds live in South America. The pink colouring comes from their crustacean foods, which they catch by swiping their rounded beak through the water.

ABOVE TOP:
**Eurasian spoonbill**
Ranging from Portugal to Japan, this species migrates down to Africa to escape colder winters in the northern parts of its range.

ABOVE BOTTOM:
**Royal spoonbill**
This Indo-Pacific species grows plumes on its head during the breeding seasons. Stick nests are built in trees beside the water.

RIGHT:
**Black-faced spoonbill**
This highly endangered species is confined to the coastlines of East Asia and breeds only on a few islands of the coast of North Korea.

ALL PHOTOGRAPHS:

**American flamingo**

Found around the southern coasts of the Caribbean Sea, this large wading bird also lives in large flocks on most of the Caribbean islands. Like all flamingoes, it has a uniquely shaped beak that allows it to feed with its head upside down, sifting microscopic food from the water.

LEFT AND ABOVE:

**Andean flamingo**

Flamingoes are able to exploit wetlands that are too harsh for other animals. The Andean flamingo feeds in highly alkaline lakes. The birds are born with white feathers, and the famed pink colouring comes from the chemicals in the food.

OVERLEAF:

**Greater flamingo**

This is the largest species of flamingo. It is widespread across the shallow lakes and coastal marshes of South Asia, the Middle East and southern Europe. During the breeding season, the whole flock will dance together, standing tall as they trot through the water displaying their wings.

**Scarlet and white ibises**
A flock of scarlet and white ibises flies together. These relatives of the spoonbills are characterised by their long hooked beaks. The white species lives further north than the scarlet one, but the two populations overlap in the wetlands of Venezuela.

# Songbirds

The bird order Passeriformes is by far the largest group of birds. At least 60 per cent of all species are in this order. A more scientific name for them is the passerines, which is derived from the Latin for 'like a sparrow'. Another term that illustrates this order is 'perching birds', which refers to the way these birds cannot really walk on the ground, but hop along instead. This because their feet and legs are mostly suited to clinging on to flimsy branches.

The more familiar term is 'songbird'. These birds have a highly evolved vocal organ, the syrinx, which allows them to produce two tones simultaneously. Of course, other birds produce calls to attract mates, find young, claim territory and send out alarms, but the songbirds – all the males and in many species the females, too – create highly musical vocalisations. The songs are incredibly intricate, too fast for the human ear to fully appreciate, and contain cultural traits – effectively regional accents – learned by listening to the calls of other birds. The song of the passerines, mostly heard at dawn before the work of foraging begins, is a way of strengthening pair bonds as well as advertising presence. Tropical passerines, especially the males, also use a wealth of bright adornments to get noticed.

OPPOSITE:
**Silver-breasted broadbill**
Somewhere in the rainforests of Southeast Asia, a parent broadbill delivers some insect food to its hungry chicks.

OPPOSITE:
**Banded broadbill**
Named for its obviously wide beak this large Southeast Asian songbird – around 20cm (8in) long – primarily hunts insects.

LEFT TOP:
**Lyrebird**
This Australian species is the largest songbird of all at almost 1m (39in) long. Spending most of its time on the ground it is named after the long harp-shaped tail plume used by the males to attract the attention of mates.

LEFT BOTTOM:
**Ovenbird**
Overwintering in Central America, this bird migrates to eastern North America to breed. It is named after its domed nests built on the ground, which resemble old-fashioned ovens.

ABOVE TOP:
**Royal flycatcher**
These insectivorous birds are found in South America. They are named after the crown-like crest that is revealed only rarely after mating. Most of the time it is neatly folded flat against the head.

ABOVE BOTTOM:
**Strange-tailed tyrant**
Sadly vulnerable to the loss of its grassland habitat, the males of this flycatcher from southern Brazil have a tail that looks like the unkempt feather belonging to a much larger bird.

RIGHT:
**Paradise flycatcher**
This paradise flycatcher from Africa is tending to its nest. It catches flying insects by snatching them from the air.

OPPOSITE LEFT TOP:
**Blue pitta**
Found throughout South and East Asia, this pitta lives in dry forests.

OPPOSITE RIGHT TOP:
**Malayan banded pitta**
As its name suggests, this species is found on the Malay Peninsula, as well as Sumatra. The 'bands' are red stripes on the breast.

OPPOSITE LEFT BOTTOM:
**Blue-winged pitta**
Living in tropical forests from India to Borneo, this species is sometimes found as far afield as Australia.

OPPOSITE RIGHT BOTTOM:
**Garnet pitta**
Named for the red breast of the male, females and young are much browner.

LEFT:
**Fairy pitta**
Breeding in East Asia and overwintering in the tropical climes of Southeast Asia, this ground-hunting bird forages for worms, insects and snails.

OPPOSITE:
**Andean cock-of-the-rock**
Known as the tunki by local people in the Andean cloud forests where this bird lives, the male cock-of-the-rock has a very distinctive quiff.

LEFT TOP:
**Fork-tailed sunbird**
Sunbirds, such as this East Asian species, are an example of convergent evolution. They live in a very similar way to the hummingbirds of America, using a long beak to access nectar from flowers.

LEFT BOTTOM:
**White-bellied sunbird**
Unlike hummingbirds, the sunbirds do not hover in front of flowers, but perch to feed. They nevertheless do use long tongues to lap up foods.

OPPOSITE:
**Loten's sunbird**
This sickle-billed sunbird lives in India and Sri Lanka.

LEFT TOP:
**Red-rumped swallow**
The swallows are an archetypal migratory bird, and this species spends winter in Central Africa and South Asia before moving north to Central Asia, Eastern Asia and Southern Europe in summer.

LEFT MIDDLE:
**Violet-green swallow**
Looking very different to the drabber swallows of the Old World, this American species migrates from Mexico as far north as Alaska.

LEFT BOTTOM:
**Rufous-bellied swallow**
This Malayan species is collecting mud for building its nest.

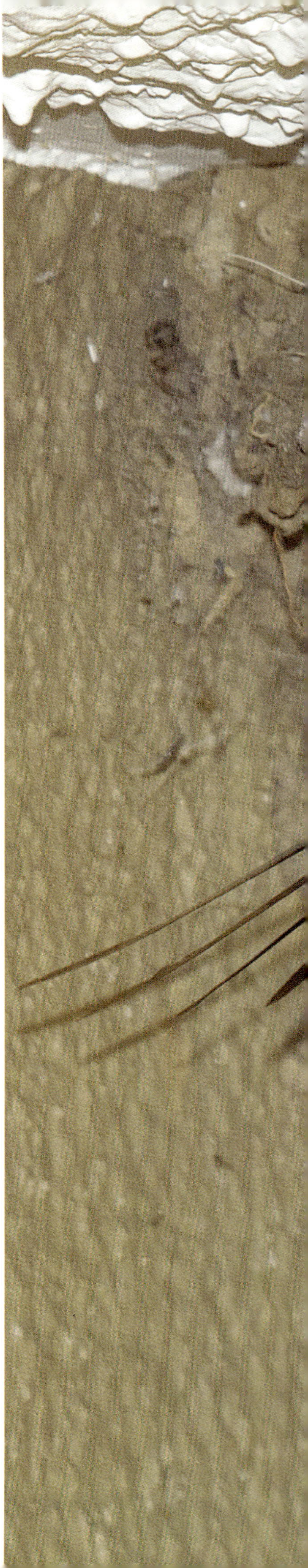

ABOVE TOP:
**Mosque swallow**
This non-migratory species lives south of the Sahara. It is named after the way it makes its nests in the mud-built mosques of the Sahel region, although it is found further afield than that, too.

ABOVE BOTTOM:
**Cliff swallows**
In summer these American swallows construct mud nests in the cliffs of the American West before heading to South America for winter.

RIGHT:
**Lesser striped swallow**
A non-migratory African species, the lesser striped swallow is known to re-use good nesting sites year after year.

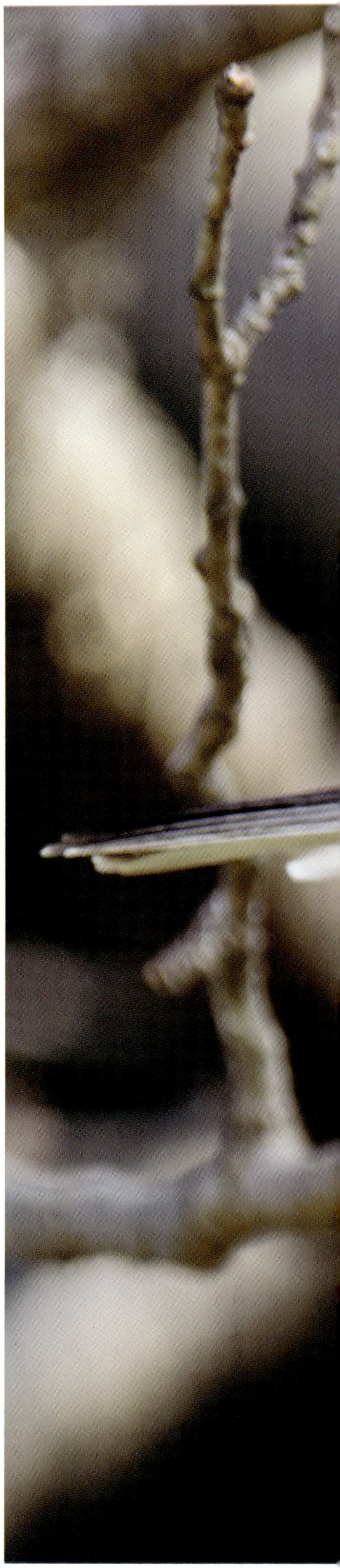

ABOVE:
**Southern white-crowned shrike**
Living in a strip of territory from Angola to Mozambique, this species lives in dry forests and grasslands. There is also a northern subspecies that lives in the savannahs of East Africa.

RIGHT:
**Bull-headed shrike**
Migrating from the south to breed in northern China, Korea and Japan, his bird perch-hunts for insects and small vertebrates.

LEFT:
**Black-throated blue warbler**
This insect-eater is supplementing its diet with a berry. It winters in the Caribbean and breeds in New England and Maritime Canada.

ABOVE:
**American yellow warbler**
The males show off a bright yellow plumage while the females are more cryptically brown. The plumage of the males varies across its range in Central America, South America and the Caribbean. Breeding grounds go as far north as Alaska.

OPPOSITE LEFT TOP:
**Tufted titmouse**
Living across eastern North America as far south as Florida and Texas, this bird has catholic tastes, eating a range of animal and plant foods.

OPPOSITE RIGHT TOP:
**Azure tit**
Mostly living from northern Asia to Finland, this non-migratory species can be found as far south as Pakistan.

OPPOSITE LEFT BOTTOM:
**Bridled titmouse**
Found in the desert woodlands of south-western USA and Mexico, this bird nests in the disused holes made by woodpeckers.

OPPOSITE RIGHT BOTTOM:
**Cinereous tit**
Once thought to be an Asian subspecies of the great tit, this bird is often seen in small flocks, gleaning insects in bushes.

LEFT TOP AND BOTTOM:
**African blue tit**
This species is found in the Canary islands and North Africa.

ABOVE TOP AND RIGHT:
**Oriental white-eye**
It's hard not to see where this Asian bird gets its name. There are also several species of white-eye across Africa and Australia, which all share the olive colouring.

ABOVE BOTTOM:
**Japanese white-eye**
Despite the name, this white-eye is found further afield in other parts of Asia, occupying mountainous regions.

RIGHT:

**Blue-faced honeyeater**

This Australian species, sometimes called the bananabird, feeds on the flowers of the banana plant but is mainly an insect-eater.

OPPOSITE:

**Eastern spinebill**

An Australian nectar-eater that uses a long beak to extract liquid food from flowers, in the same way that hummingbirds and sunbirds do.

LEFT:
**Golden-breasted bunting**
A striking little African bird that hunts for spiders and insects on the ground. It lives in dry wooded areas south of the Sahara.

ABOVE TOP:
**Yellowhammer**
This widespread Eurasian bird moves to the hot Middle East in winter.

ABOVE BOTTOM:
**Lark-like bunting**
A nondescript bunting that uses its drab colours to blend into the dusty semi-deserts of south-western Africa.

OPPOSITE:
**Green-headed tanager**
This colourful finch-like bird lives in the Atlantic forests of Brazil.

LEFT TOP:
**European goldfinch**
An all-year resident of western Europe, this beautiful seed-eater heads south to the Middle East and North Africa after spending summer in Eastern Europe.

LEFT MIDDLE:
**Purple finch**
One of the American rosefinches, this species prefers to winter around the Gulf of Mexico and the Deep South.

LEFT BOTTOM:
**Eurasian bullfinch**
In winter this bird often moves to warmer regions in the south, around the Mediterranean and Central Asia.

OPPOSITE:

**Black-headed weavers**

A female of this African weaver looks cosy in her nest, woven from strands of grass.

LEFT:

**Asian golden weaver**

The bright yellow males of this species attract mates by building nest balls hanging from trees. The birds nest in colonies.

ABOVE:

**Venezuelan troupial**

The common name of this bird is derived from the French word 'troupe', the root of the English term 'troop'. It refers to the way these birds live in large flocks.

RIGHT:

**Golden oriole**

Named after the vibrant yellows of the male (which is seen in flashes on the female plumage, too), the orioles are famous for building well-constructed nests in the forks of branches.

PREVIOUS PAGE LEFT:

**Mayotte drongo**

The drongos are tropical relatives of the crows. This species is found only on a small island in the Indian Ocean.

PREVIOUS PAGE RIGHT:

**Satin bowerbird**

This is a female of this Australian species with greenish feathers, similar to its relative, the catbird. The male is a glossy black.

OPPOSITE:

**Flame bowerbird**

Bowerbirds are named after the elaborate twig structures the males construct to woo females. This species is from New Guinea.

LEFT TOP:

**Regent bowerbird**

The male of this Australian species has a distinctive golden mantle on its back and head.

LEFT BOTTOM:

**Green catbird**

This bowerbird is from Australia's Queensland coast.

FAR LEFT:
**Magnificent bird of paradise**
Birds of paradise show off their amazing feathers during elaborate displays. In between, they eat fruits and insects.

LEFT TOP:
**Raggiana bird of paradise**
The birds of paradise of New Guinea have taken eye-catching plumage to the extreme. The group name comes from the mistaken belief that these birds had no feet and never landed – like an angel. This idea arose because early specimens returned to Europe by explorers had had their legs removed.

LEFT MIDDLE:
**Superb bird of paradise**
Only the males invest in these elaborate adornments, which hinder flight. The fact that a male thrives despite his cumbersome costume is a signal to females that he is a healthy mate.

LEFT BOTTOM:
**Twelve-wired bird of paradise**
This species is 33cm (13in) long, a medium size for a bird of paradise. There are 42 species, which also go by the names riflebird, manucode, parotia and sicklebill.

OVERLEAF LEFT:
**Greater bird of paradise**
At 43cm (17in) long, this is the largest bird of paradise. The males gather in small groups called 'leks' to display to females.

OVERLEAF RIGHT:
**Western parotia**
This mountain species performs a dance with its feathers spread out as a skirt.

# Picture Credits

**Alamy:** 8 (Ray Wilson), 19 (Nature Picture Library), 22 (Imagebroker), 23 (Sylvain Oliviera), 32 (Hira Punjabi), 37 (Lee Dalton), 40 (Image Professionals), 44 top (Imagebroker), 50 left (Danita Delimont Creative), 54 bottom (Tom Ingram), 57 (Prisma by Dukas Pressagentur), 60 top (Gary & Donna Brewer), 66 (Imagebroker), 67 top (Marco Valentini), 67 middle (Minden Pictures), 81 (Gabbro), 100 bottom left (Neil Bowman), 105 (Watters Wildlife Photography), 121 (Kevin Schafer), 124 bottom (Agami Photo Agency), 127 (Raju Ahmed), 128 (Arterra Picture Library), 129 top (Minden Pictures), 131 left (All Canada Photos), 132 (Sipa US), 215 (Buiten-Beeld), 218 (Nature Picture Library), 221 (Corbin17), 221 bottom (David Tipling Photo Library)

**Dreamstime:** 10 bottom (Dreamstime5752), 17 bottom (Beeboy3), 20 (Smellme), 25 (Dudlajzov), 30 (Phichak Limprasutr), 31 top left (Dennis Jacobsen), 31 top right (Slowmotiongli), 31 bottom (Lee Dalton), 36 (Nicole Duperre), 41 top (Stanislav Duben), 41 middle (Dennis Jacobsen), 44 bottom (Tamara Bauer), 50 top right (Patrice Correia), 50 bottom right (Agami Photo Agency), 51 bottom (Thawats), 52 (Ray Shiu), 53 top (Kittisak Terananon), 53 middle (Richard Mcmillin), 55 (Dennis Donohue), 56 (Artushfoto), 58 (Ray Shiu), 61 (Edurivero), 63 (Mikelane45), 67 bottom (Agami Photo Agency), 68 (Scratchart), 69 bottom (Noor Adimah Muslim), 70 (Rupam Mandal), 71 top (Lim Or Or), 71 bottom (Shi Hao Liao), 72 (Rinus Baak), 73 top (Brian Lasenby), 73 bottom (Jana Scigelova), 74 (Wildphotos), 75 (Harry Collins), 84 top left (John Platt), 84 top right (Panuruangjan), 88 bottom (Nikolai Sorokin), 90 (Brian Kushner), 92 bottom (Gerrit De Vries), 94 (Mhpiper), 95 bottom (Neal Cooper), 97 bottom (Bokgallery), 102/103 (Satheesh Rajh Rajagopalan), 107 (Klomsky), 111 middle (Patrice Correia), 112 top (Steve Allen), 112 bottom (Vladimir Cech), 114 (Tony Bosse), 120 bottom (Jiri Hrebicek), 122/123 (Riondt), 136 (Hellmann1), 140 bottom (Ekays), 141 (Miroslav Liska), 142 (Adogslifephoto), 143 (Bernhard Richter), 145 (Ktree), 146 bottom (Wrangel), 148/149 (Simon Fletcher), 151 (Rinus Baak), 152 top left (Suttipon), 152 top right (Panuruangjan), 155 top (Tratong), 155 middle (Liqiang Wang), 158 (Kajornyot), 160 (Robert Downer), 161 top (Kathy Alsop), 161 bottom (Sharon Jones), 162 (Brian Grant), 163 top (Donyanedomam), 164/165 (Wirestock), 166 (Moose Henderson), 167 (Gerald D Tang), 168 top (Marcus Scott-parkin), 168 middle (Lukas Blazek), 168 bottom (Ajay Bhaskar), 169 (Lebedinski), 170/171 (Lukas Blazek), 172/173 (Donyanedomam), 174 bottom (Martin Pelanek), 175 (Liqiang Wang), 177 top (Andrey Gudkov), 177 bottom (Wrangel), 178 (Delstudio), 184 (Chuchart Duangdaw), 188 bottom (Gabriel Rojo), 190 top right (Thawats), 190 bottom left (Kajornyot), 190 bottom right (Feathercollector), 191 (Panuruangjan), 192 (Manik1770), 193 top (Wenjunchu), 194 (Malcolmarouza), 195 top (Lee Amery), 198 (Arno Meintjes), 199 (Raju Soni), 200 (Philip Rathner), 201 (Karen Burgess), 202 top left (Wildphotos), 202 bottom left (Rinus Baak), 203 top (Leif Ingvarsson), 203 bottom (Victor Suarez Naranjo), 204 top (Chuchart Duangdaw), 206 & 207 (Paddyman2013), 208 (Martin Pelanek), 209 top (Andrew Allport), 209 bottom (Altaoosthuizen), 210 (Mikelane45), 211 top (Sindlera), 211 middle (Ron Hoff), 211 (Stig Karlsson), 212 (Chris Van Lennep), 213 (Phichak Limprasutr), 214 (Ben Schonewille), 217 (Andreas Ruhz), 219 top (Meg Forbes), 219 bottom (Tanya Puntti), 220 (Liewwk), 221 top (Feathercollector), 222 (Omar Ariff Kamarul Ariffin), 223 (Liewwk)

**Dreamstime/ Ondrej Prosicky:** 45, 62 bottom, 80, 101, 113, 131 bottom right, 134/135, 180/181

**Getty Images:** 129 bottom (Education Images)

**Shutterstock:** 6 (DG Stock), 7 (Karel Bartik), 10 top (Jesus Cobaleda), 11 (Coulanges), 12, 13 (Danny Ye), 14/15 (Andrii Slonchak), 16 (Olivia Oates), 17 top (Joost van Uffelen), 18 (Teekayu), 21 (Ondrej Prosicky), 24 (Alec Trusler2015), 26/27 (Uliya Krakos), 28 (Nomadic Exposure), 29 bottom (David Steele), 33 top (Rafael Martos Martins), 33 bottom (Gabriela Insuratelu), 34/35 (Anan Kaewkhammul), 38/39 (Sevde Sevan), 41 bottom (worldswildlifewonders), 42 (Narupon Nimpaiboon), 43 top (Alpio), 43 bottom (Itsuky), 46 (Kristan Bell), 47 top (John Carnemolla), 47 bottom (Trevor Scouten), 48 top (Barbara Ash), 48 bottom left (Jayden Gunn), 48 bottom right (Ken Griffiths), 49 (RL Hambley), 51 top (Marcos Amend), 53 bottom (pittawut), 54 top (Joseph Fernando), 60 bottom (Milan Zygmunt), 62 top (Martin Pelanek), 64/65 (Miguel Lincango), 69 top (Butterfly Hunter), 76 (Jean-Edouard Rozey), 77 (Jennifer Bosvert), 78 (RaiPhoto), 79 top (guentermanaus), 79 bottom (noicherrybeans), 82 (kajornyot wildlife photography), 84 bottom (Dariush M), 85 (Ondrej Prosicky), 86/87 (Hit1912), 88 top (Karel Bartik), 88 middle (luis m Valadao), 89 top (Riza Marlon), 89 bottom (teekayu), 91 (Carol Ann Moyse), 92 top (Andre Marais), 93 (Butterly Hunter), 95 top (Dave Montreuil), 96 (Terry Dell), 97 top (feathercollector), 98 (Ken Griffiths), 99 (LouieLea), 100 top left (Martin Mecnarowski), 100 top right (Agami Photo Agency), 100 bottom right (Tomas Drahos), 104 top (Luka Hercigonja), 104 bottom (Nick Scheltema), 106 (Super Prin), 108/109 (Radek Borovka), 110 (Super Prin), 111 top (Mada-Foto), 111 bottom (Bildagentur Zoonar), 115 top (Nico Fourie), 115 middle (Chris Sheppard), 115 bottom (Tomas Calle Boyero), 116 (Milan Zygmunt), 118/119 (Ondrej Prosicky), 120 top (Jiri Hrebicek), 124 top (Rafael Martos Martins), 125 (Paul Wittet), 126 (punkbirdr), 130 (Rob Jansen), 131 top right (tee262), 133 (Pablo Rodriguez Merkel), 138 (Nikolay Hristakiev), 139 top (Kensho Photographic), 139 bottom left (Honza123), 139 bottom right (Holger Kirk), 140 top (Matt Deakin), 144 (Sergei25), 146 top (Rudi Ernst), 147 (meunierd), 150 top (Super Prin), 150 bottom (Volodymyr Burdiak), 152 bottom right (Michel Sloviak), 154 (Heying HUA), 156 (mooinblack), 157 top (Sunti), 157 bottom (feathercollector), 159 (Neat Soup), 163 bottom (JMxImages), 174 top (Dilomski), 176 (Nadezda Boltaca), 177 middle (Gudkov Andrey), 179 (Tony Mills), 182/183 (Michael Mescheryakov), 186 (BGS_Image), 187 top (Andreas Ruhz), 187 bottom (Frode Jacobsen), 188 top (Tiny Turkey), 189 (Boonchob Chuaynum), 190 top left (Super Prin), 193 bottom (Martin Mecnarowski), 195 middle (David Spates), 195 bottom (Muhamad Mizan bin Ngateni), 196 top (Dave Montreuil), 196 bottom (Jeff Huth), 197 (Lynne Wilde), 202 top right (Anton MirMar), 202 bottom right (Zens Prarom), 204 bottom (Dave Hansche), 205 (Tareq Uddin Ahmed), 216 (Agami Photo Agency)

**Shutterstock/ Wang LiQiang:** 29 top, 71 middle, 152 bottom left, 153, 155 bottom, 157 middle